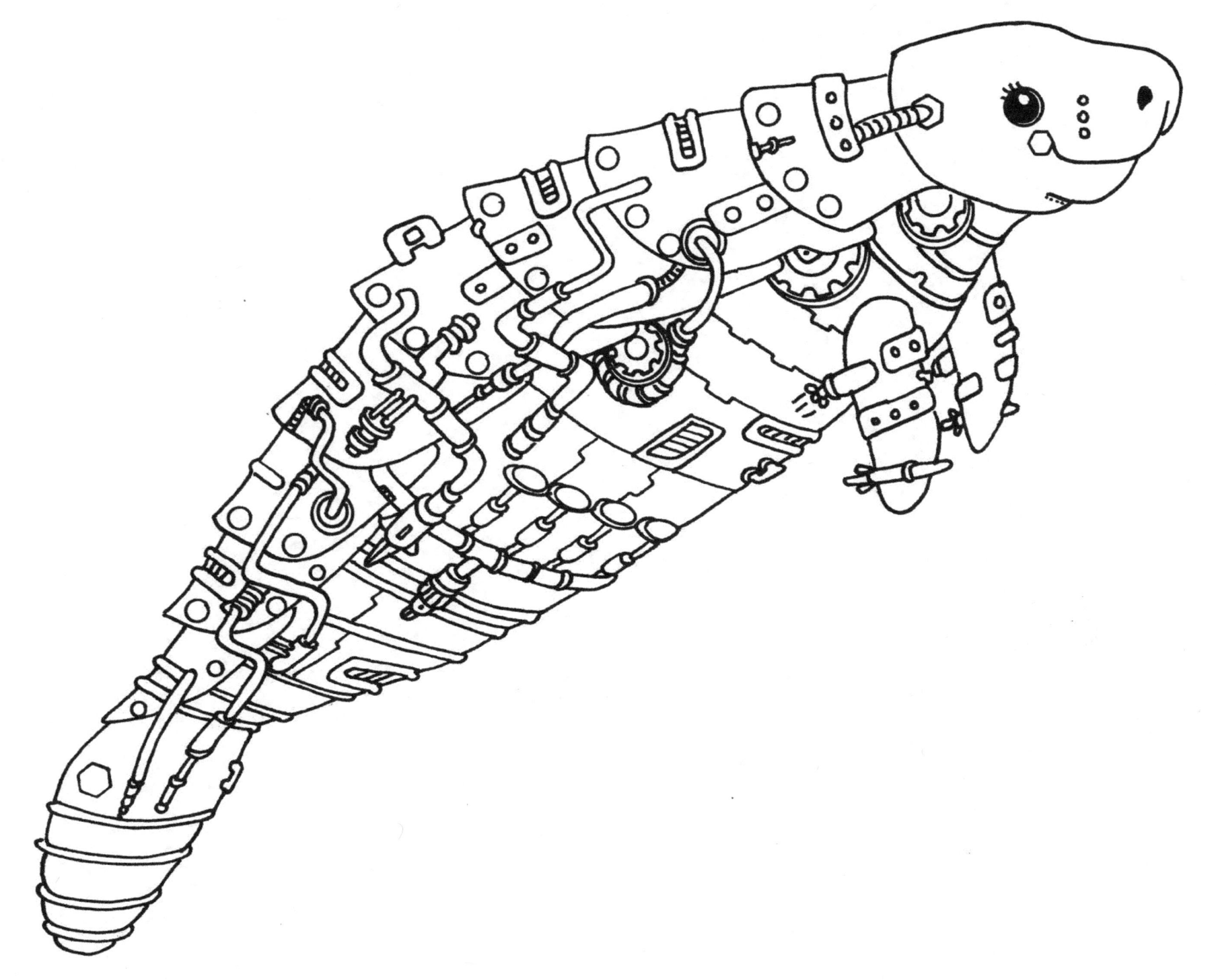

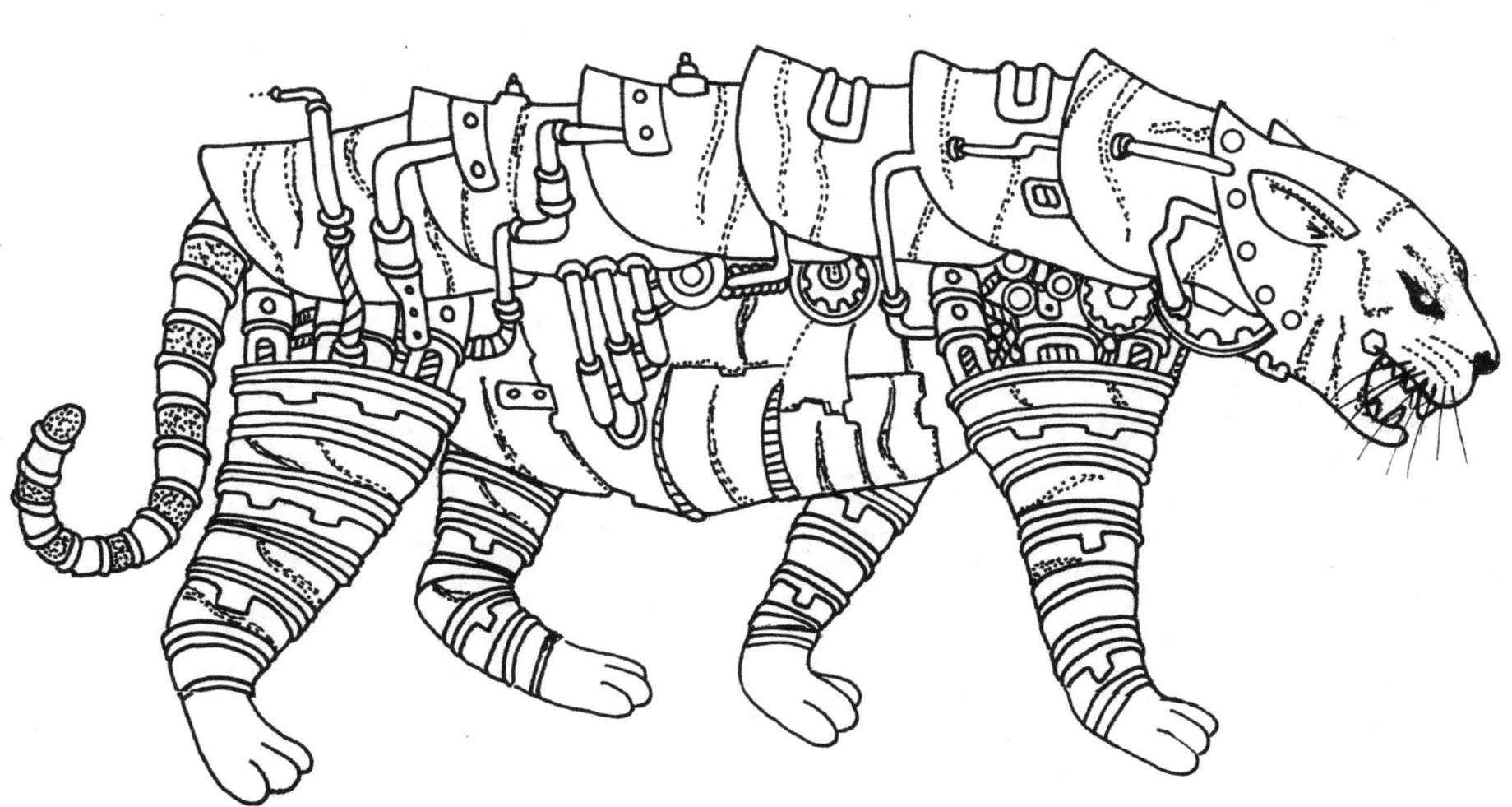

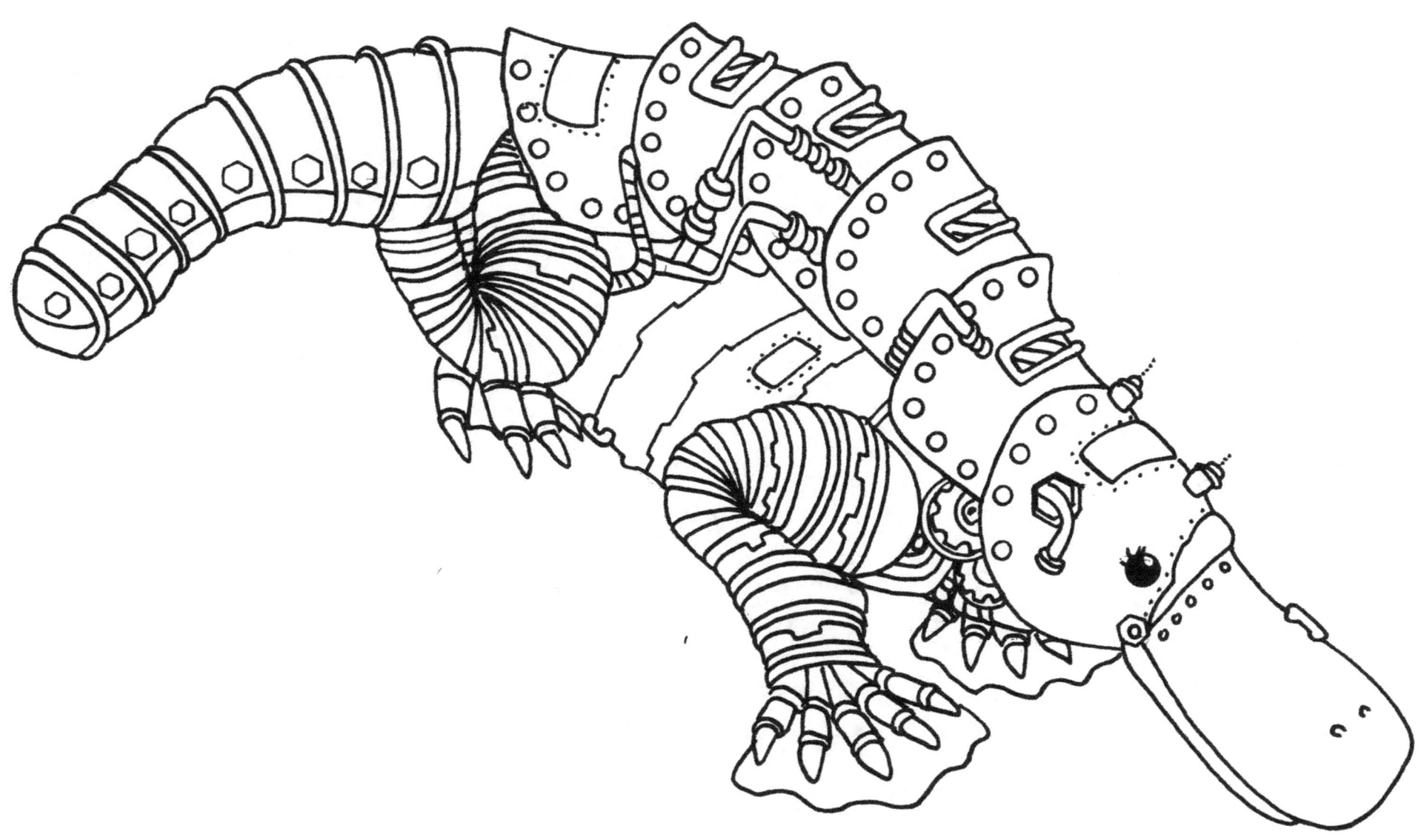

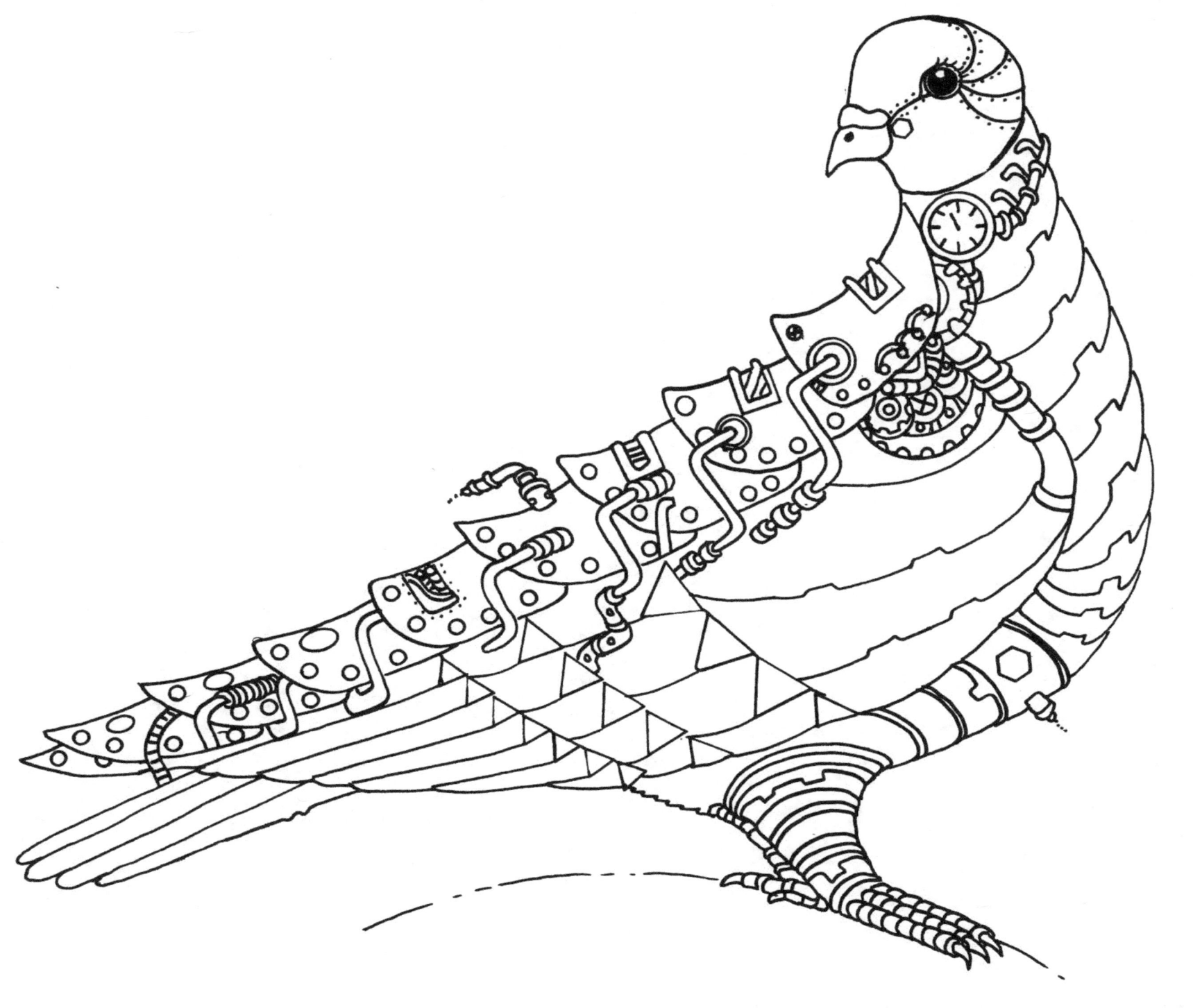

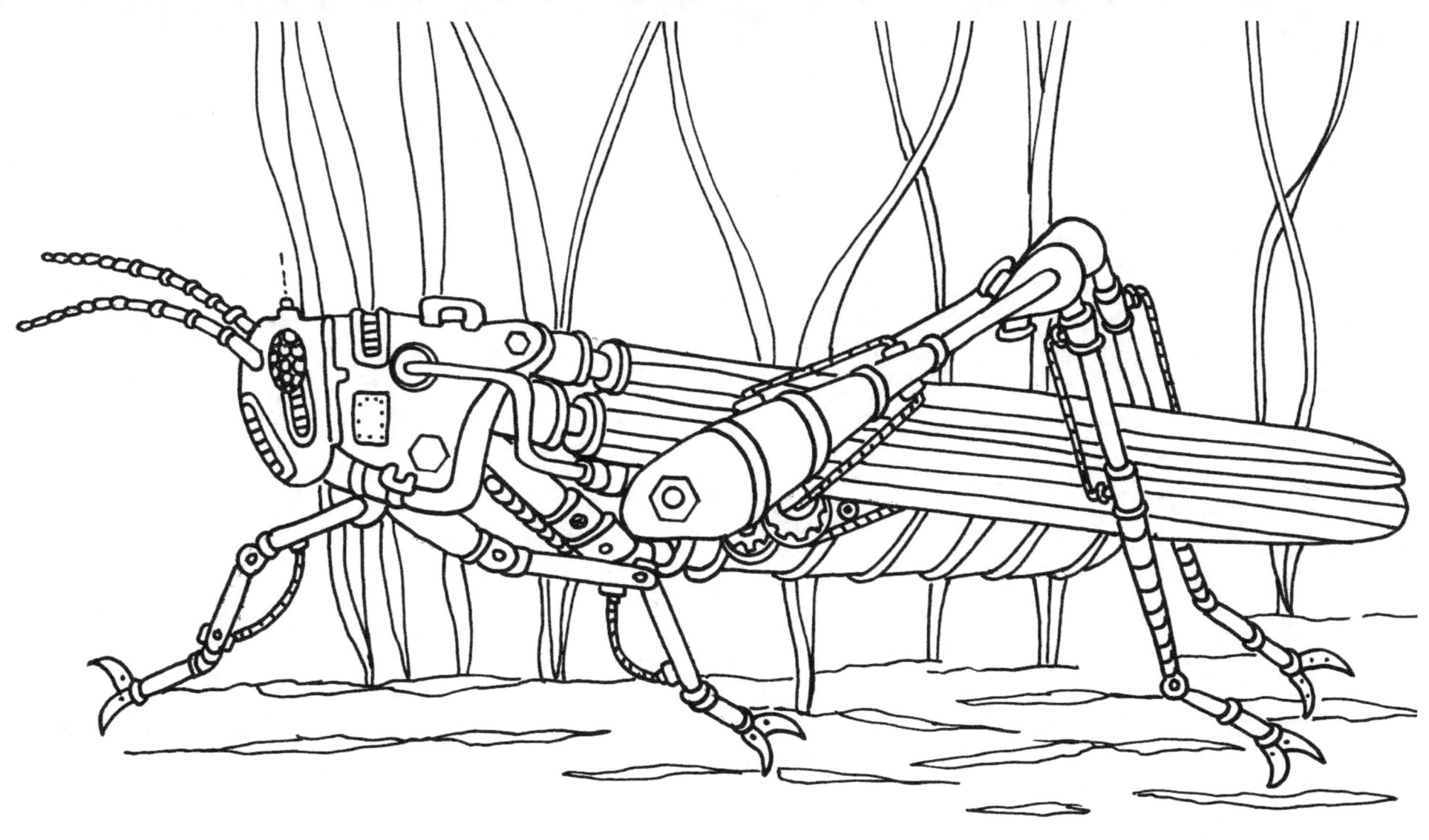

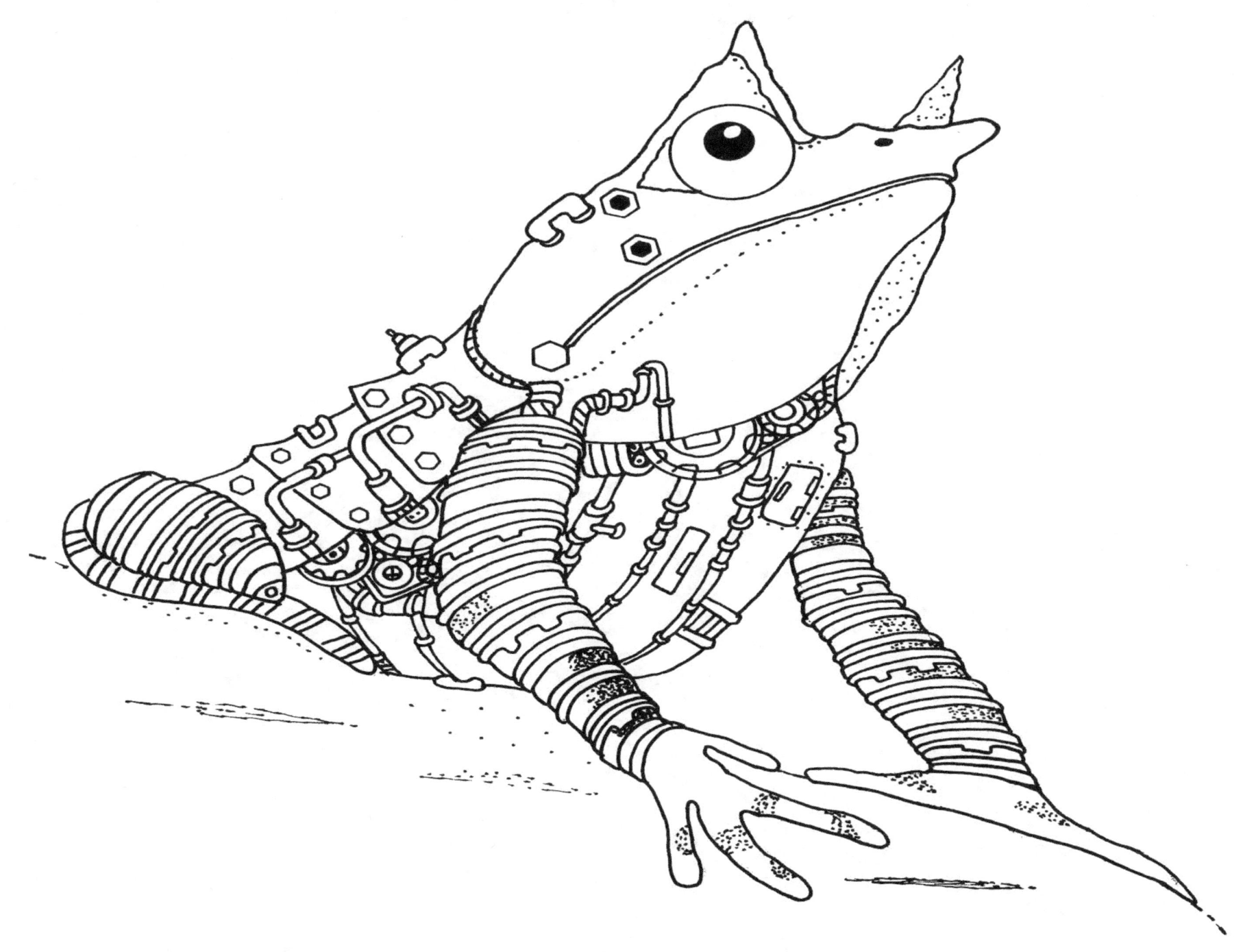

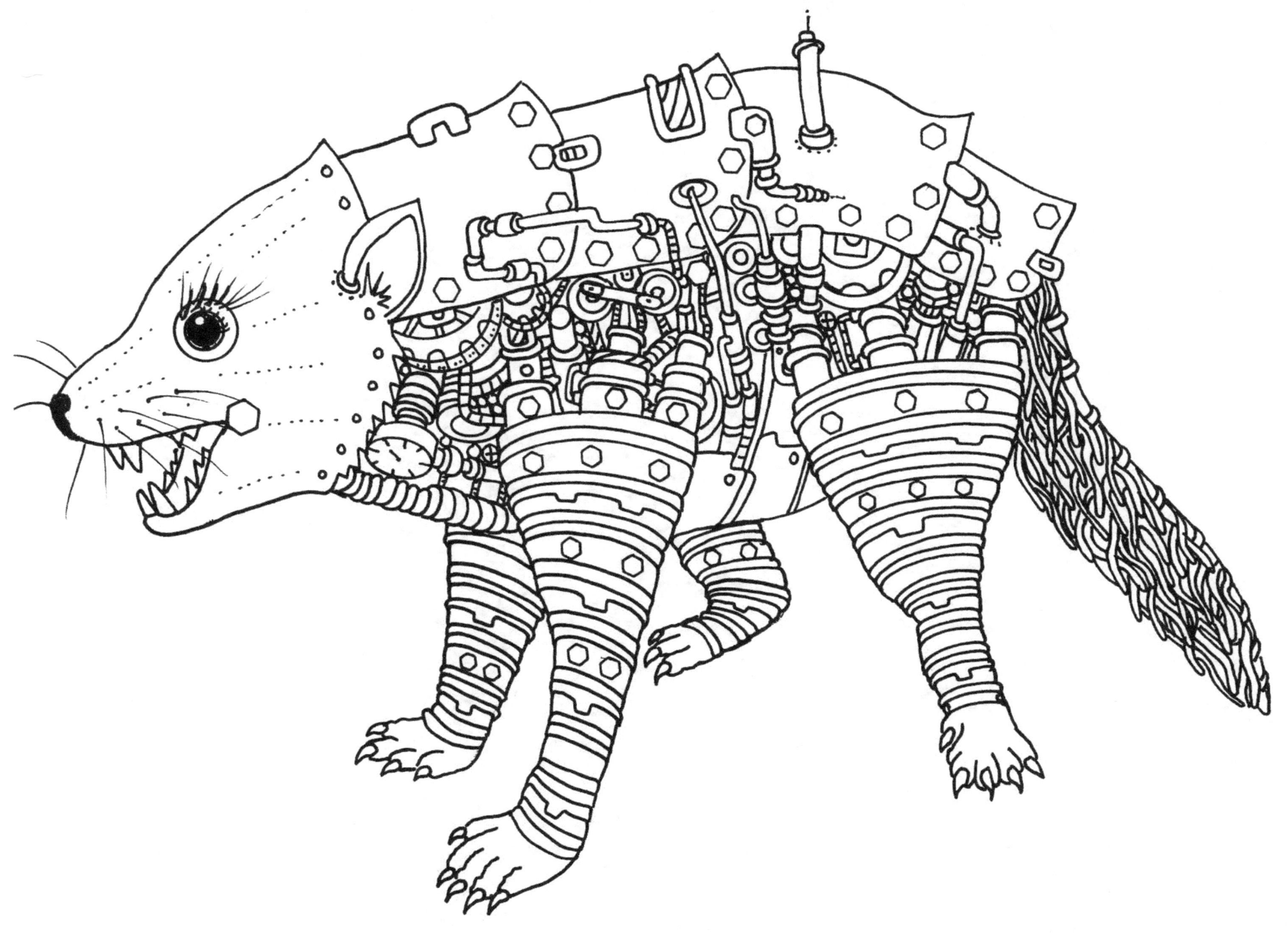

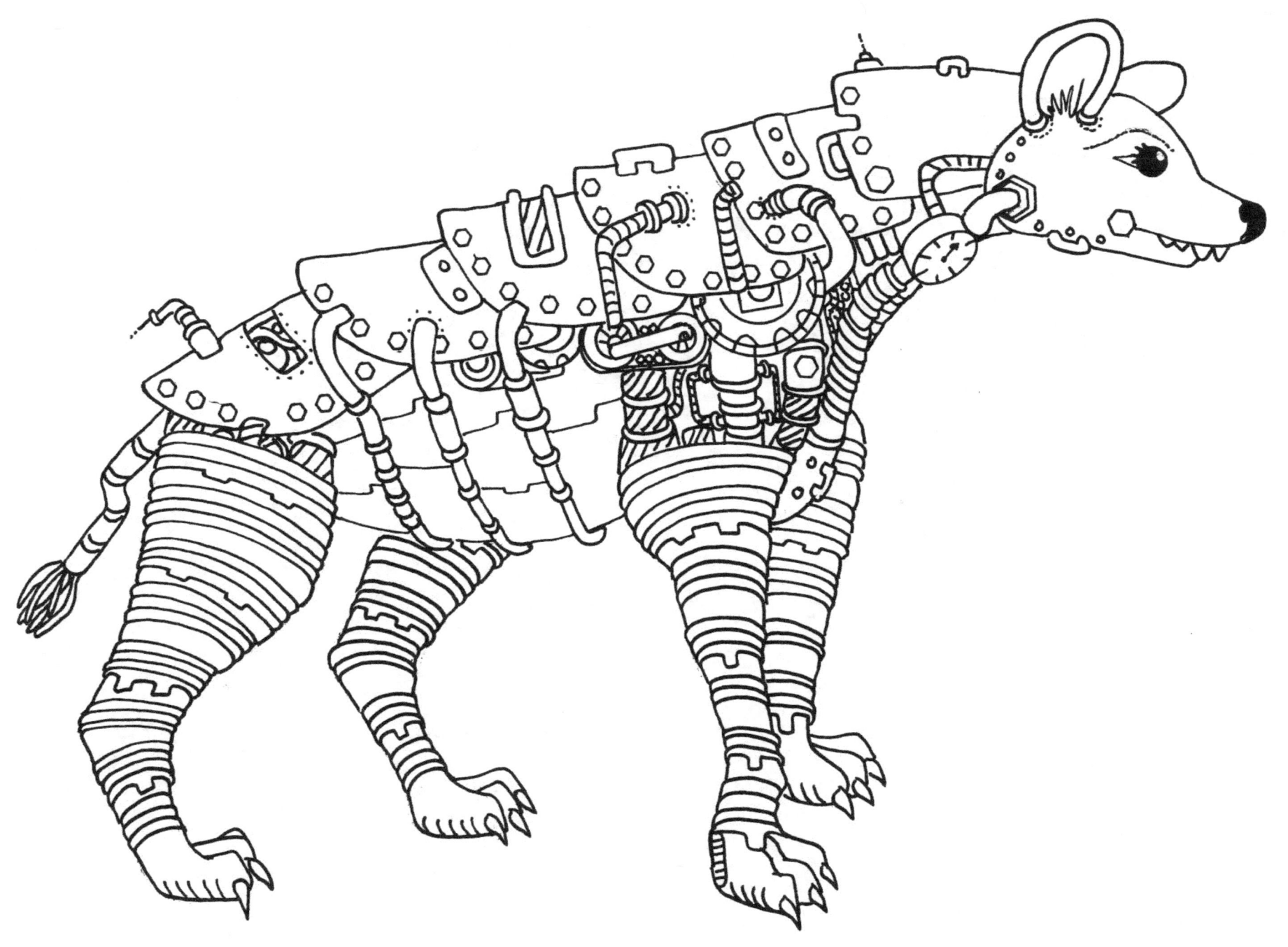

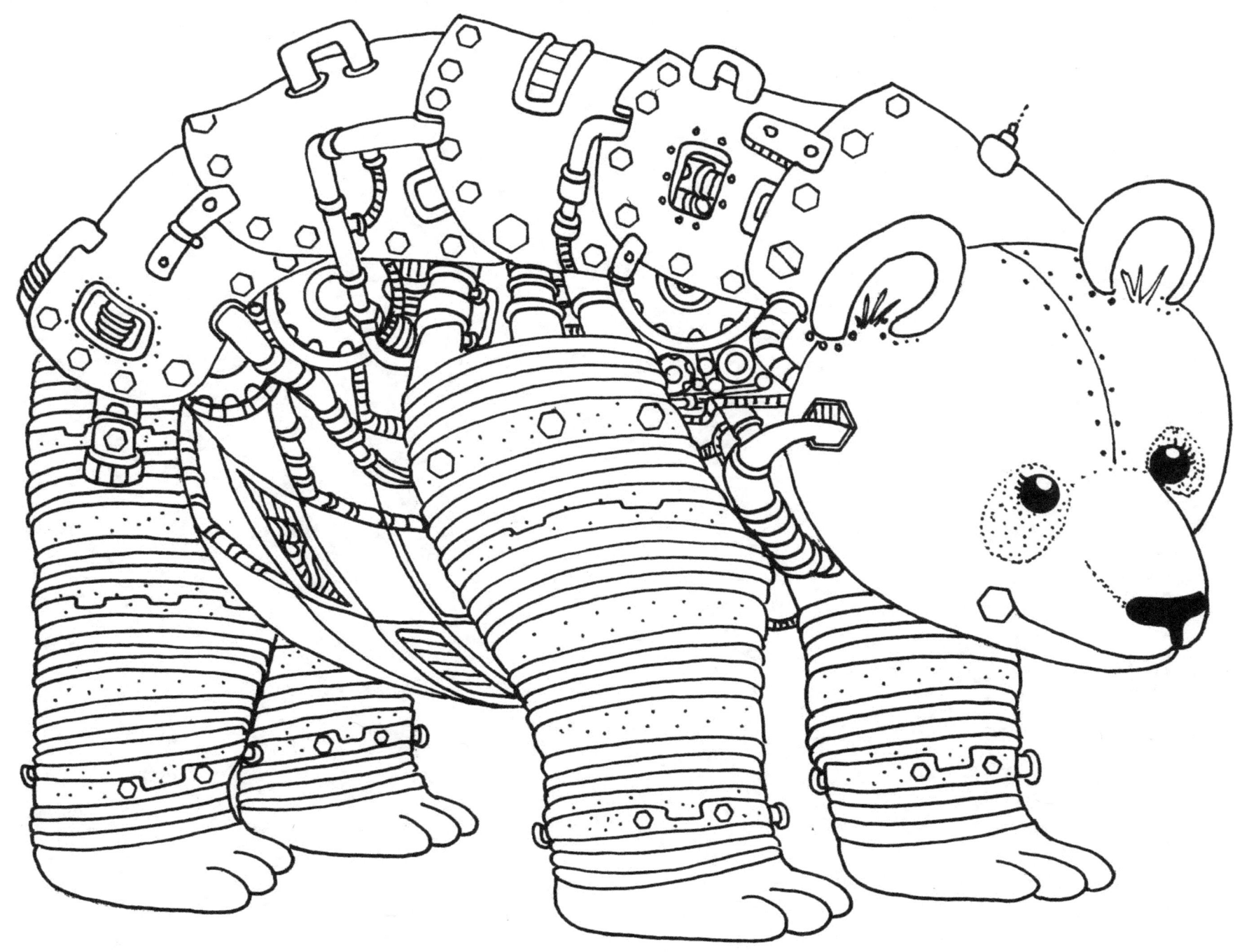

MECHANICAL CRITTERS

Volume Four of Mechanical Critters:

If you enjoy this coloring book, check out volumes 1-3 of Mechanical Critters. This exciting coloring book introduces you to twenty five of the world's most well known critters.

Simply add color to this gallery of critters to create imaginary images of a Polar Bear, Dromedary, European Badger, Golden Retriever(Pointer), Opossum, Grasshopper, Kangaroo Rat, Malaysian Horned Frog, Red Fox, Manatee, Seal, Zebra, Raccoon, Tiger, Indian Cow, Pigeon, Platypus, Young Female Lion, Tasmanian Devil, Wolverine, Elk, Short Tailed Mole, Panda, Hyena, and a Crocodile. These critters are hand drawn with a fine point black marker and a lot of imagination in a mechanical form, using nuts, bolts, gears, sprockets, pipes, armored plating, tubing, and many other mechanical parts.

These incredible illustrations are original images created
by a well-known artist and art educator, Timothy L. Worachek.

Purchasing this and any of my other coloring books will provide continued inspiration and support through my retirement. Each volume contains twenty-five different critters. Volume Four contains the 100th critter illustration so far.

I would like to publish your colored Mechanical Critters in up coming volumes. Please send a scanned image, saved as a jpeg, or PDF, with the first name, age, location(State or Country) and comments to: timothylworachek@gmail.com

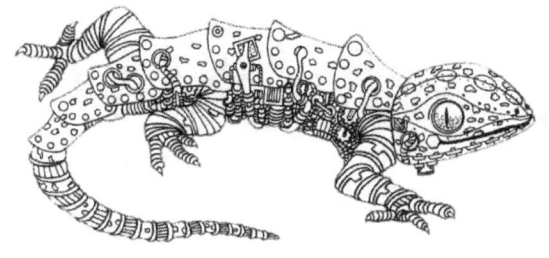

www.ingramcontent.com/pod-product-compliance
Lightning Source LLC
Chambersburg PA
CBHW080138240526
45468CB00009BA/2518